W0044188

Your INTELLIGENT BRAIN

Dr Mike Tranter

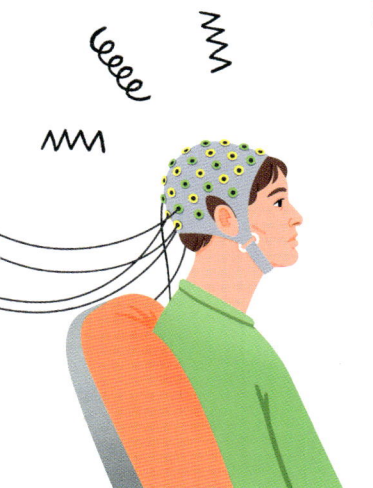

OXFORD
UNIVERSITY PRESS

Great Clarendon Street, Oxford OX2 6DP

Oxford is a registered trade mark of
Oxford University Press in the UK and in certain other countries

© Oxford University Press 2024
Text written by Dr Mike Tranter
Illustrated by Geraldine Sy and Ana Seixas

Designed and edited by Raspberry Books Ltd

British Library Cataloguing in Publication Data:

ISBN 978-0-19-278314-1

1 3 5 7 9 10 8 6 4 2

Printed in China

Paper used in the production of this book is a natural,
recyclable product made from wood grown in sustainable forests.
The manufacturing process conforms to the environmental regulations
of the country of origin.

Acknowledgements

The publisher and authors would like to
thank the following for permission to use
photographs and other copyright material:

Cover artwork: Geraldine Sy and
Ana Seixas; Oxford University Press
photos: Pavlo S/Shutterstock and
author. **Inside artwork:** Photos:
p7(bm): AnyaPL/Shutterstock; p7(br):
Zhenyokot/Shutterstock; p7(t): Mackey
Creations/Shutterstock; p8: Jose Luis
Calvo/Shutterstock; p31: rachaya
Roekdeethaweesab/Shutterstock; p32: Denis
Production/Shutterstock; p33: Creactivomx/
Shutterstock; p45: Mike Goldsmith; p54:
wavebreak media/Shutterstock; p55:
GoodFocused/Shutterstock; p56: Ljupco

smokovski/Shutterstock; p67: © Public
Domain; p70: rawisoot/Shutterstock; p71:
DedMityay/Shutterstock; p72: Semnic/
Shutterstock; p77: Gilmanshin/Shutterstock;
p88: StudioPhotoDflorez/Shutterstock;
p88(b): Pixel-shot/Shutterstock.

Artwork by **Geraldine Sy**, **Ana Seixas**,
Aaron Cushley, Raspberry Books and
Oxford University Press.

Every effort has been made to contact
copyright holders of material reproduced
in this book. Any omissions will be rectified
in subsequent printings if notice is given to
the publisher.

Images are to be used only within the context of the pages in this book

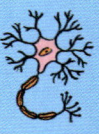

Contents

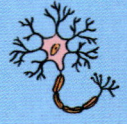
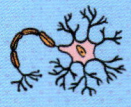
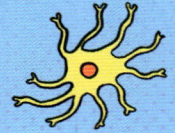

Chapter 1

What is the Brain?

Sitting securely inside your head, the brain is the most
essential part of your body. It is where your thoughts,
feelings, memories, and personality come from.
It also processes everything your senses detect,
helping you understand the world around you.

Your brain is filled with specialized **cells** that are very
good at quickly processing information. Think of it
as the **control centre** for almost everything
you do—like having a big computer in your head.
It constantly communicates with your body, telling
it how to move, breathe, eat, and sleep.

If you could look **inside your head** right now,
you would see a soft, spongy, dark-pink brain that is
jelly-like and slimy. It weighs about 1.4 kg (the same
as seven oranges) and its pink colour comes from
its vast blood supply. It needs a lot of blood because
blood carries oxygen and glucose (sugar) that your
brain uses for energy.

A brain has many folds called gyri. Gyri increase the brain's surface area, allowing it to fit in more brain cells, and giving humans more **processing power**. There are lots of grooves in the brain, called sulci, which look a little like wrinkles.

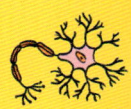

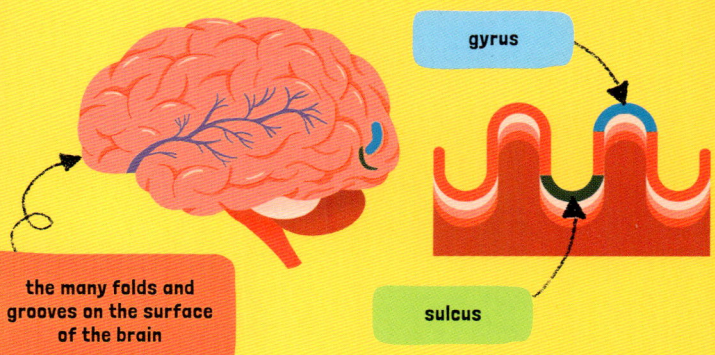

gyrus

the many folds and grooves on the surface of the brain

sulcus

✳ **Sp**ea**k** like a **sc**ie**nt**ist ✳

GYRI AND SULCI

Gyri (say jai-rai) are the folded, raised areas of the brain. A single fold is called a gyrus. Sulci (say sulk-ai) are the deep grooves that separate the brain. A single groove is called a sulcus. Together, these create space for more brain cells and processing power.

Hearts and minds

In ancient times, people thought that it was the heart that did everything. That was until an ancient Greek scholar, Alcmaeon of Croton, came along around the year 500 BCE. He was the first person to understand that it was the brain doing all of our complicated thinking, not the heart.

BRAIN HERO

ALCMAEON OF CROTON

Ancient Greek scholar who was the first person to work out that the brain does our thinking.

Does the size of the brain matter?

Humans have the most advanced brain in all of the animal kingdom. It is divided into sections that perform different activities (see pages 26–33). One important feature is the size of our brain relative to our body. An elephant may have a brain three times the size of ours, but although it is intelligent, it is not as intelligent as a human.

BRAIN SIZE AND WEIGHT

sperm whale:
9 kg (the biggest on the planet, nearly 6.5 times the size and weight of the human brain)

African elephant:
4.6 kg (nearly 3 times the size and weight of the human brain)

human:
1.4 kg

stegosaurus:
0.070 kg or 70 g (smaller than a tennis ball; around 20 times less than the weight of the human brain)

In this very short introduction to the amazing brain you'll discover:

- that one human has around **86 billion** brain cells called **neurons**

- what's travelling along the brain's **information superhighway**

- how **smelling** a pizza influences its taste

- why Sir Isaac Newton pushed a **needle** behind his own eyeball

- how the brain can be tricked into **seeing** something . . . but not into tickling ourselves.

Prepare to have your **mind blown** as we explore **the body's supercomputer.**

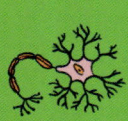

Chapter 2

What is the Brain Made Of?

The brain is a collection of cells that communicate with each other to complete tasks.

Of all the different types of brain cells, none are as crucial as the neuron. The neuron is the brain cell responsible for your thoughts, feelings, movements, and everything else you can think of. It forms connections with other neurons that branch out to create networks and circuits with different parts of the brain.

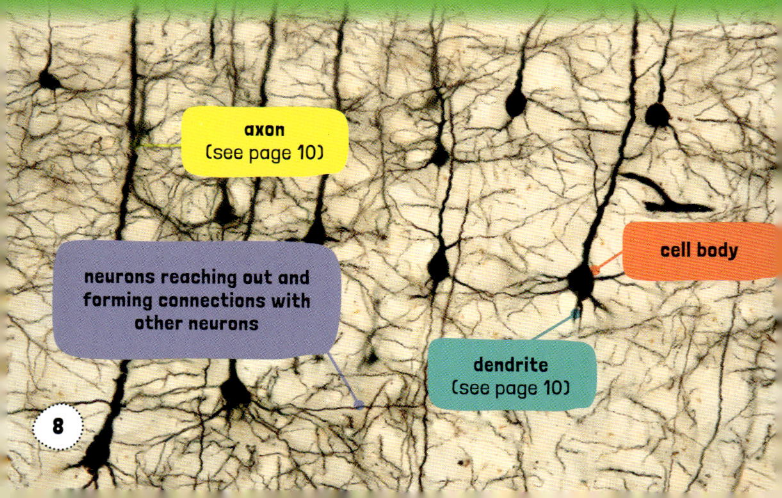

axon
(see page 10)

cell body

neurons reaching out and forming connections with other neurons

dendrite
(see page 10)

Speak like a scientist

NEURON

A neuron is a type of cell found in the brain and body which sends signals to control everything we think and do. A bundle of neurons is called a nerve.

Neurons can then send signals to neurons in other parts of the brain. What is most impressive is that one single neuron can have up to **10,000 connections!**

The brain contains an extraordinary number of neurons. We know this because scientists decided to slice up some brains and reduced them down to their bare minimum components in a sort of brain soup. Then they added a special dye to turn the neurons into a bright colour to make them easier to count. They counted brain cells in all of the different brain regions and realized that the brain contained around **86 billion neurons.** That's a lot of counting when you think that 86 billion would be ten times as many people as there are on Earth today.

A neuron is made up of a cell body, which is a little like a head, containing the nucleus that holds the **DNA** instructions for how each cell works. This is where the hair-like dendrites are too, receiving signals from other neurons. It also has an axon, the long 'arm' of the neuron which the signal travels along. When the signal gets to the end, **communication** with other neurons can occur.

Neurons communicate by sending out chemicals, called **neurotransmitters**, from their endings, or 'axon terminals'. The synapse is the gap between one neuron and another, and it is where the communication happens. The synapse sits between the axon terminal of one neuron and the dendrite of a second neuron. A neurotransmitter is **30,000 times smaller** than the thickness of a human hair and the whole process of communication takes only two milliseconds; that's **500 times faster** than one second!

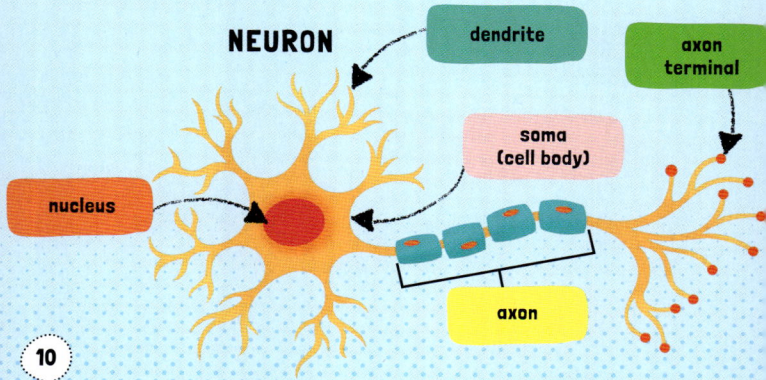

NEURON

dendrite

axon terminal

soma (cell body)

nucleus

axon

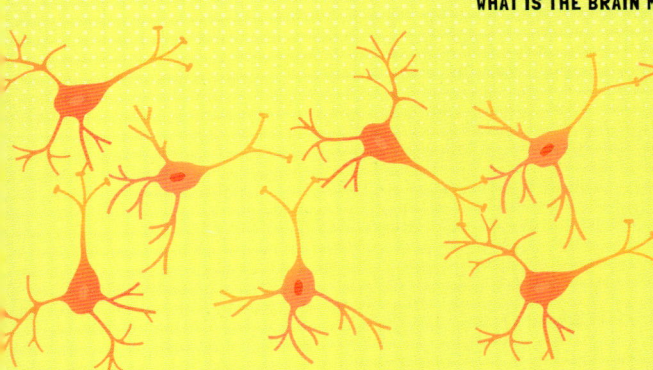

✳ **Sp**ea**k** like a **sc**ie**nt**ist ✳

NEUROTRANSMITTERS AND SYNAPSES

A neurotransmitter is a chemical molecule that passes from one neuron to another.

A synapse is the gap between one neuron and another. Neurotransmitters move across this gap.

There are many types of neurotransmitters helping neurons to do specific things (see pages 12–13). It is a little bit like **language**. Some neurons communicate in specific languages and only neurons which also know those languages will know how to respond.

Neurotransmitters in the brain

Neurotransmitters are intricately involved in many processes throughout the brain. Here are some of the most important ones and what they do:

DOPAMINE: Makes us feel good, motivates us to do things, and also helps us to learn. It is crucial in how our brain tells our body to move.

SEROTONIN: Makes us feel happy and influences our mood, helps to control sleep and regulates our appetite so we don't feel too hungry or crave excessive amounts of food.

GLUTAMATE: Activates other neurons and tells them to keep firing signals; important in forming long-term memories and decision-making but also significant in many different brain functions.

GABA: Stops the neurons from sending messages, and helps to control brain signals and the flow of communication in the brain.

NORADRENALINE: Helps us to wake up after sleeping, increases our attention, and keeps us alert, especially when problem-solving.

Now time for some homework!

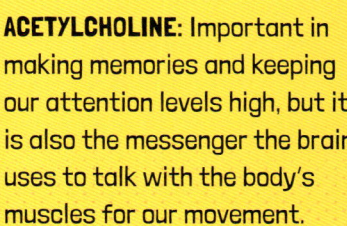

HISTAMINE: Helps to keep us awake so we don't fall asleep during the day, helps to motivate us to work towards our goals, and stops us from feeling hungry.

ACETYLCHOLINE: Important in making memories and keeping our attention levels high, but it is also the messenger the brain uses to talk with the body's muscles for our movement.

Along with neurons, your brain has other types of cells too. Microglia act as the brain's immune system, fighting off infections and controlling any inflammation.

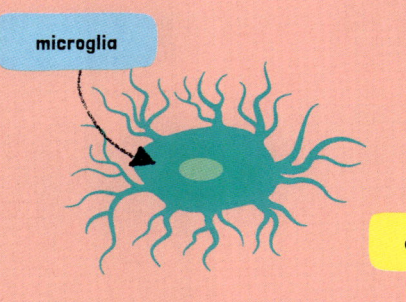

microglia

astrocyte

Astrocytes support neurons to function by recycling neurotransmitters, and help neurons to grow and develop properly.

Brain matter

If we could **slice** through the brain, we would see two distinct layers of grey and white. Unsurprisingly, we call this grey and white matter. It is a general way of grouping all neurons in the brain and explaining where they are. If we sliced through, we would see the colour change from pinky-grey towards the surface to pinky-white in the middle.

About 40% of the brain is made up of grey matter, which comprises the cell bodies of neurons. What is amazing about the brain is that when we are younger, it makes more brain cells than we need. Over time, it learns which ones to keep and reduces the overall number. Technically, children have bigger brains than fully grown adults.

A CROSS-SECTION OF THE BRAIN SHOWING THE LAYERS OF GREY AND WHITE MATTER

grey matter, neuronal cell bodies

white matter, neuronal axons

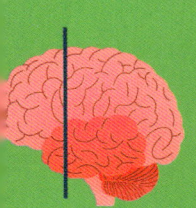

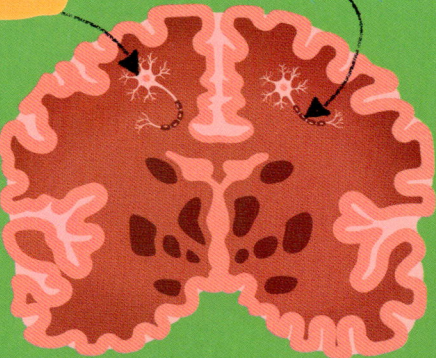

White matter makes up the other 60% of the brain and contains the neurons' long, stretched-out axons. These axons are like the arms of the neuron, reaching out to thousands of other neurons, **sending signals** to make your brain do complex things.

Axons in the brain vary in length, from around 1 mm to several centimetres, and you need a lot of them to connect everything together in the brain. If we tied together all the axons of a single human, they would stretch to well over 500,000 km in length. That could get you all the way **to the Moon!**

Axons are wrapped in a fatty material called **myelin**, created by a special cell in the brain called the oligodendrocyte. The myelin helps the neurons to send their signals more efficiently. Think of it as a little like pouring water down a slide before you use it—you will slide down more easily (and it will be much more fun too).

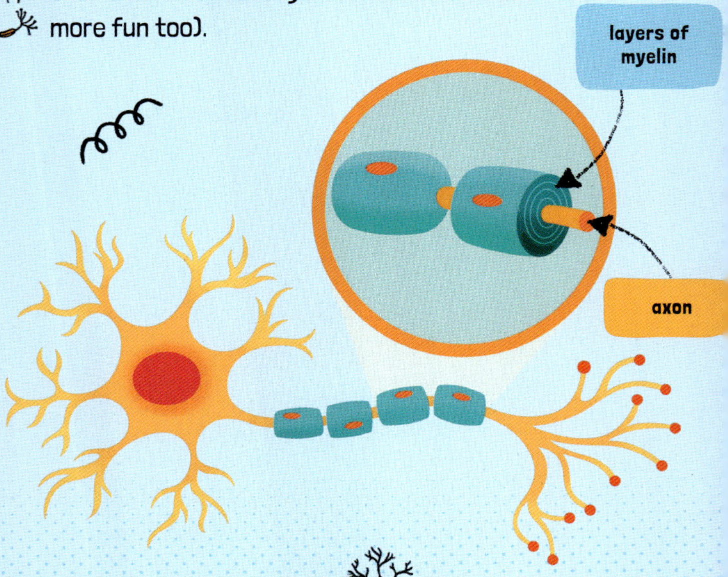

layers of myelin

axon

Everything we think or do is because of how neurons communicate through their signals, called **'action potentials'**. An action potential is really a wave of electrical charge produced by **ions** (atoms with more or fewer electrons than they would normally have) moving in and out of the axon very quickly. The action potential can travel along the axon at speeds of over 400 km per hour.

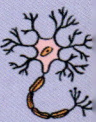

✳ Speak like a scientist ✳

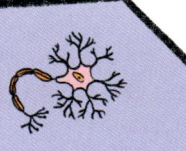

ACTION POTENTIAL

Action potential is the name given to the signal, or electrical charge, that travels along a neuron's axon.

In the next chapter, we'll see what happens in the brain when all of these cells work together to influence everything that the body does, and how the brain has specialized features, like the **blood-brain barrier** (see pages 20–21), **to help keep it safe.**

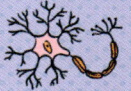

Chapter 3

How Does the Brain Work?

The brain contains billions of neurons, all sending signals and communicating with each other to perform certain tasks.

If the neurons in the brain want to send instructions to the body, they need to send signals to the neurons in the spinal cord. The spinal cord contains **nerves** that send instructions from the brain to the body, such as how to move your arm to reach for something. It also sends messages back into the brain.

✱ Speak like a scientist ✱

SPINAL CORD

The spinal cord is like a superhighway for all communication between your brain and body. It is a long tube of tissue running all the way down your back, containing bundles of neurons (nerves).

The brainstem

The brainstem is an immensely important region that merges with spinal cord neurons between the brain and the body. It regulates your body temperature, controls your sleep, helps you move, affects your blood pressure, tells your heart to keep beating and your lungs to keep breathing. Basically, the brainstem performs many of the automatic things in the body that keep you **alive.**

brainstem

It is also responsible for things you might never have thought about before, like vomiting, coughing, and sneezing. That's your brainstem at work trying to get rid of something that is potentially harmful.

The thalamus

thalamus

The **thalamus** is an egg-shaped region at the top of the brainstem. It is essential for relaying signals from neurons involved in our movement or senses. We understand a lot about the thalamus thanks to the French scientist, Cécile Vogt (see page 20).

BRAIN HERO

CÉCILE VOGT

French neuroscientist who mapped out the human brain and provided the basis for our understanding of how brain regions communicate with each other.

The cranial nerves

The **cranial nerves** consist of twelve pairs of nerves that come out of the brain. They are in pairs because each half of the brain has its own set. Mainly, they control muscles in the face and sense when it is touched, but they are also responsible for things like eye movement, smell, and taste. They tell the muscles in our face or mouth to move, helping us to make facial expressions or chew our food. If you have ever eaten something very cold very quickly, the headache or **'brain freeze'** you feel is the result of one of these cranial nerves (the trigeminal nerve) being activated.

The blood-brain barrier

Your brain needs protection from harmful things reaching it. Bacteria, viruses, and many drugs are stopped from getting into the brain by something called the blood-brain barrier.

Speak like a scientist

BLOOD-BRAIN BARRIER

The blood-brain barrier is the layer of cells lining blood vessels that protect the brain from harmful things in the blood.

CAPILLARY CROSS SECTION

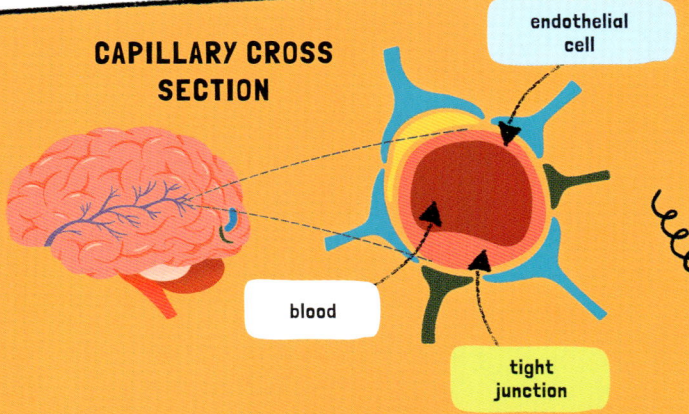

endothelial cell

blood

tight junction

The blood vessels that deliver blood to the brain have a particular type of cell lining them, called **endothelial cells**. These cells work together with tight junctions, which act a little like the cement between the bricks in a wall, creating a barrier between the blood and the brain. This barrier says **'yes'** to things that are good for your brain, like glucose, and **'no'** to all the bad stuff, like bacteria and viruses.

Cerebrospinal fluid

Your brain is currently **floating** in a clear liquid called the cerebrospinal fluid, or CSF. It acts as a protective layer, helping to minimize injuries whilst also removing waste products from the brain, such as damaged cells and excess molecules it doesn't need any more. It also contains nutrients that your brain needs, like salts, sugars, and proteins. Doctors can sometimes test the CSF to check brain health and look for signs of infection.

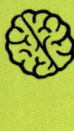

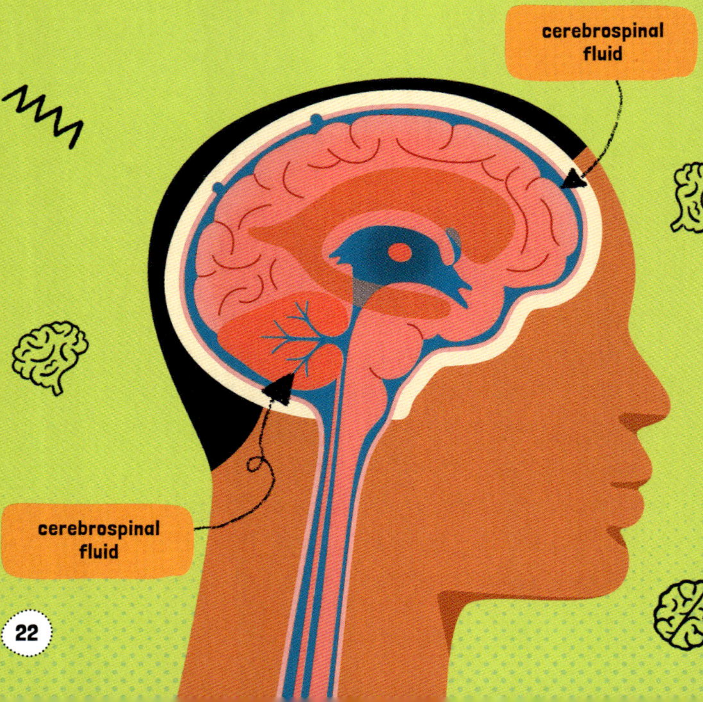

cerebrospinal
fluid

cerebrospinal
fluid

The CSF flows through the spinal cord and ventricles of the brain, which are like tunnels for the CSF to pass through. There is around 120 ml of CSF in your brain right now. This increases to 150 ml in adults—that's the same volume of about half a can of soft drink. If that doesn't sound like a lot, that's OK, because new CSF is made continuously throughout the day and night to keep it really fresh. When we sleep, the flow of CSF is washing through our brains **in rhythmic waves**.

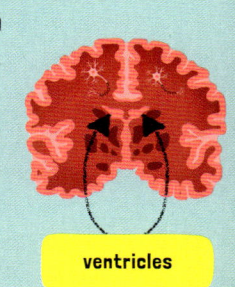

ventricles

 Speak like a **scientist**

CEREBROSPINAL FLUID

Cerebrospinal fluid (CSF) is the clear fluid that cushions the brain and spinal cord and provides nutrients to the brain.

Now we understand what the brain is made of, in the next chapter we will look at how it is organized so that **all its different regions work together effectively**.

Chapter 4

How is the Brain Organized?

The entire brain is split into two parts called hemispheres. Each hemisphere has a number of regions called lobes.

Looking at the brain from the top, you can see each **hemisphere** clearly, with a ridge in between them called the **longitudinal fissure**. They are joined together by the **corpus callosum**, which helps the two hemispheres communicate and work together. The corpus callosum is a bundle of neurons that transmit information between hemispheres. In fact, there are more than 200 million neurons here—that's a lot of information!

THE BRAIN FROM ABOVE

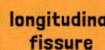

longitudinal fissure

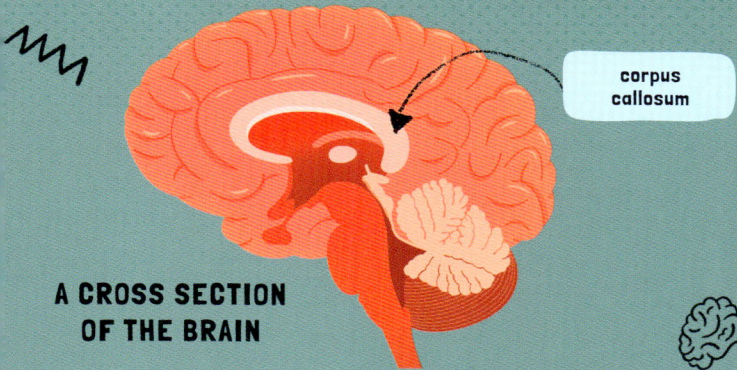

corpus callosum

A CROSS SECTION OF THE BRAIN

A **fascinating** feature of the brain is that the right hemisphere controls the left side of the body, and the left hemisphere controls the right side. When neurons leave one side of the brain and travel down the spinal cord into the body to tell it what to do, they actually have to cross over to the other side. When messages travel back towards the brain, they cross back over.

Our brains have specific areas responsible for different types of tasks. In fact, there are around 200 smaller regions that our brains can be divided into. Along with the brainstem, the brain is organized into five major regions called **lobes**. For example, the temporal lobe is important in memory, and the occipital lobe in vision.

Let's take a look at the main lobes in more detail.

THE LOBES OF THE BRAIN

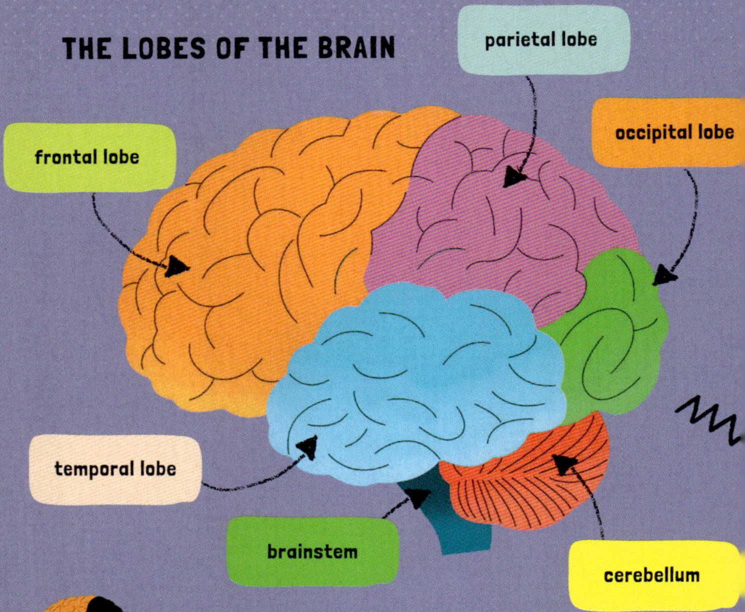

parietal lobe

occipital lobe

frontal lobe

temporal lobe

brainstem

cerebellum

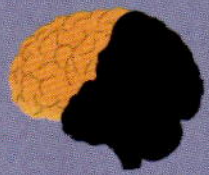

The frontal lobe

The frontal lobe sits behind the forehead and it is where much of our personality comes from. We understand so much about the frontal lobe because of **groundbreaking research** in the late 20th century by an American neuroscientist called Patricia Goldman-Rakic. She helped us to understand how the brain develops, and how it can manage certain types of injuries. She dedicated a lot of her research to understanding what is happening in the brain cells and lobes when we think and make decisions.

Speak like a **scientist**

FRONTAL LOBE

The frontal lobe is the part of the brain behind the forehead. It is essential in planning, decision-making, and logical thinking.

The frontal lobe likes to think ahead and plays a big role in planning and making decisions. It considers any potential consequences of our actions and helps guide us into making the best decision in the moment. Our decision-making improves as we get older, partly from experience and partly because our frontal lobe develops over time, until we are at least twenty-five years old.

BRAIN HERO

PATRICIA GOLDMAN-RAKIC

American neuroscientist who led the research on the function and development of the frontal lobe.

The parietal lobe

The parietal lobe is critical for things like language, writing, and maths. It is also essential for receiving and interpreting signals from the body about physical touch and temperature.

The parietal lobe plays a big part in helping us process information about the world around us and where our body is at any given moment. It also assesses information from the senses, like taste and smell, which we will learn about in the next chapter.

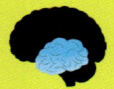

The temporal lobe

The temporal lobe is known for learning and memory as it contains a small seahorse-shaped region called the hippocampus, which is needed for short-term memory. If you remember something from about sixty seconds ago, that

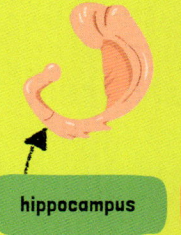

hippocampus   **seahorse**

information has been sitting in your hippocampus waiting to be used. The temporal lobe is also involved in language (although lots of brain regions are needed for this) and is really active when listening to music.

Speak like a scientist

HIPPOCAMPUS

The hippocampus is a structure in the brain that we need for our memories, especially short-term ones.

Deep within the temporal lobe there is a group of structures collectively called the basal ganglia. This area is responsible for telling us how to move, and for sending those instructions to our body.

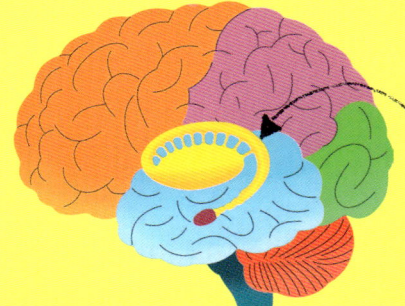

basal ganglia

Another function of the temporal lobe is recognizing the faces of people we know. There is a particular region at the bottom of the brain called the fusiform face area which is responsible for this.

29

The occipital lobe

The occipital lobe is all about vision. Signals from the eyes travel along the optic nerve to the thalamus before signalling to a location in the occipital lobe, which is at the back of the head. The image can then be passed to other areas of the brain, with each one giving us a clearer understanding of what we are looking at and whether it is familiar (see page 34).

One of the many exciting things about this region is that it is always active, even when you are sleeping. That's why there are visual elements to our dreams.

Sometimes, when people close their eyes tightly, they can see colours. These are called phosphenes and occur even though our eyes are closed, because the retina at the back of the eye (a layer of light-sensitive cells) is still sending lots of signals. With our eyes closed, there isn't really much to see, so we only get flashes of light or colour.

HOW IMAGES TRAVEL TO THE OCCIPITAL LOBE

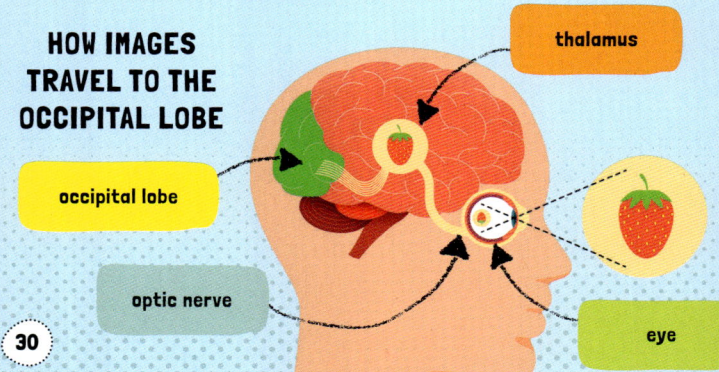

thalamus

occipital lobe

optic nerve

eye

Newton and phosphenes

The famous mathematician Sir Isaac Newton was also interested in phosphenes. He discovered that light is made up of lots of colours and observed his own phosphenes by pushing a needle just behind his own eye to stimulate his retina. Not only is this disgusting but it is **extremely dangerous** and should never be attempted by anyone!

BRAIN HERO

SIR ISAAC NEWTON

English physicist and mathematician famous for his work explaining gravity.

Thankfully, today we have unique cameras to look at the eye without the need for painful and dangerous tactics like Newton's. His sketches of these experiments have been preserved and can be seen today.

The cerebellum

The cerebellum is best known for helping humans and other mammals with balance and coordination. When a baby learns how to sit up, their cerebellum figures out how to keep their balance. If they manage to stay upright, the cerebellum remembers what they did and repeats it. It is also very important in what are called 'fine motor movements', such as picking up a pencil or playing a musical instrument.

Want to know something hilarious?

Have you ever tried to tickle yourself? It doesn't work, does it? Even if we are ticklish, we can't seem to tickle ourselves.

Being tickled can get uncomfortable quickly because it causes an intense feeling of uneasiness and stress.

But rather than run away or scream, we laugh.

The thing is, we can't send ourselves into that stressful state because no matter how hard we try, our brain always knows that we are safe and can stop at any moment. This reaction is controlled by our cerebellum and parietal lobe, which know the difference between a sensation that you are expecting (tickling yourself) and one that is surprising to you (being tickled).

Our brains work hard to decode and interpret all the information around us, not just who is about to tickle us.

In the next chapter, we are going to take a look at exactly how the brain interprets the world around us through our senses, **and how we can really put them to the test**.

Chapter 5

Super Senses

One of the most important of the brain's jobs is
processing all of the information around us—what
we touch, taste, smell, hear, or see.

Let's take a look at the **primary senses** and
see how our brains use the information they give us
to work out what is happening around us.

Sight

For people to see anything, light needs to reflect
off an image, enter the eyes, and hit the retinas
(see pages 30 and 37). The retina forms a layer at the
back of each eye and turns the light into an electrical
signal. This signal travels along the optic nerve all the
way to the visual cortex in the occipital lobe.

The brain needs to process information about what
we are looking at—colour, size, movement, distance—
and tell our brains what it is. Even if only the vague
shape of something can be made out, the brain can
eventually learn to recognize it. Memory regions in
the brain try to remember it, the frontal lobe helps us

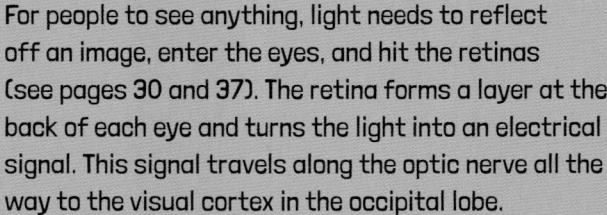

to make a decision about what to do, and the brain's emotional regions tell us how we feel.

Let's take a look at how it works:

Your eyes see an animal running towards you and send the images to your brain via the optic nerve.

Your visual cortex and the memory areas of your brain will identify its shape and size, recognize what it is, and you notice it getting closer.

Crocodile!

The emotional areas of your brain tell you how you feel.

I feel scared!

Your frontal lobe tells you what to do.

I need to get away!

Phew!

Optical illusions

Things happen very quickly in the real world and, as impressive as the brain is, it still takes a little bit of time to process all of the information the eye sends to it. This means that the brain has to improvise and predict what it expects to see, then match it with what it actually sees. Optical illusions trick the brain into making incorrect predictions.

For example, take Kanizsa's triangle. The brain generally expects to see a triangle because there is an empty space forming that shape. But all that is actually in the picture are some circles with missing parts and some lines. An imaginary triangle is only seen because the brain is being tricked—it is an illusion.

KANIZSA'S TRIANGLE

The triangle in the middle doesn't really exist, but the eye is tricked into seeing it.

The scintillating grid illusion

This optical illusion is made up of a grid of white lines
on a dark background. Where one line meets another,
there are white circles. This contrast of light and dark
can trick the brain into seeing flashing black circles
which are never really there. Scintillating comes from
a Latin word meaning 'to sparkle'.

This is how the scintillating grid illusion works. When
looking directly at an image, the light goes evenly
into the eye and hits the middle of the retina. In this
spot, the eye generally sees very clearly, but if light
enters from the side—peripheral vision—the retina
doesn't see as clearly and needs to adjust itself to
the brightness of what it sees. But this adjustment
also changes the rest of the vision. When looking at the
scintillating grid, the cells of the retina are constantly
adjusting themselves to make up for seeing so many
dots. Some are clear and some are not, and the
brightness changes with that clarity, making it seem
like dots are flashing all around the page.

Some readers can be very sensitive to these types of
optical illusions, so we haven't shown the scintillating
grid here, but it can be found online.

When two eyes work together, they generally enable someone to see things in **three dimensions,** which makes assessing distance easier. For example, if someone reaches for a glass of water on a table, when both of their eyes work together it helps them determine how far away it is, so they don't end up knocking it on to the floor.

Someone might be able to catch a ball with one hand, but with one eye closed, they will probably find it difficult to judge the distance.

The brain adapts to many situations and, for those used to not seeing in three dimensions, this task will be much easier. For those who aren't used to it, it will get easier with time and practice.

The time it takes for the signal to travel to the visual cortex and to be processed is around 0.1 seconds, creating a short delay.

This means when the ball travels in the air, at least some of what is seen is based on the predicted trajectory of the ball. If the brain didn't predict this movement, essentially guessing where it is going to be, then the small time delay would mean the ball might hit the person before they can catch it.

Taste

How do our brains know that the delicious scoop of ice cream tastes so good? The taste buds on our tongues have many **receptors** that send signals to our brains about what we are eating.

taste bud

The chemicals that give food its flavour bind to those receptors. The signals travel through cranial nerves to the gustatory cortex in the brain where the taste is perceived.

Smell

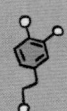

Our sense of smell works on the same principles as taste. Essentially, chemicals travel into the nose, where they bind to receptors, which lead to neurons sending signals to the brain.

From there, sensory neurons project to an area of the brain called the olfactory bulb (connected to the olfactory cortex), where information about the smell is processed.

This information is linked with the rest of the brain to determine what to do about it. For example, if the smell was pleasant, we might feel happy about that because it could mean **delicious** food, or disgusted if it was a **terrible** smell of rotten food. The brain is suspicious of things that smell or taste bad because they could be harmful to us, and makes us disgusted. Because of this, neurons that interpret smell are linked with our **limbic system**. The limbic system is a term for multiple brain regions, which are responsible for generating our emotions.

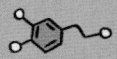

HOW THE BRAIN PROCESSES SMELLS

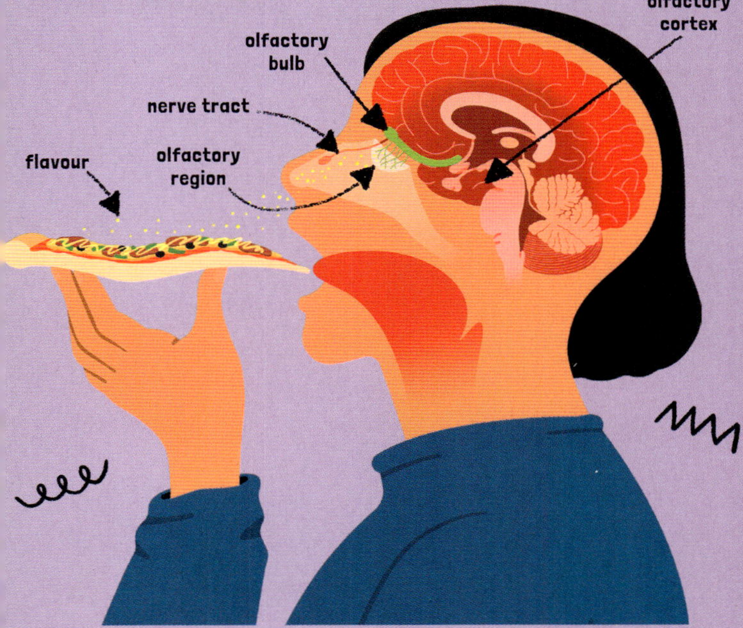

olfactory cortex

olfactory bulb

nerve tract

flavour

olfactory region

Our sense of smell is important in another way too. Smells can help us to taste things because receptors in the nose detect smells and tell the brain about them in the olfactory cortex. The brain processes information from the taste buds and the nose before finally deciding on the taste. Of course, we also get information about texture, temperature, and remembering eating particular foods, which all **help our experience.**

Experiment time!

Try it out yourself. **Hold your nose** when you eat (make sure you can still breathe through your mouth) and see how your food tastes. You may find it much more difficult to taste things without using your nose, showing you just how much of an impact your sense of smell has when eating. This is because what you eat creates a smell that manages to travel to your nose as you are eating, and together with your tongue, creates a taste. For an even bigger challenge, close your eyes so you can't see what you're eating.

The Proust effect

Have you ever smelled something that sparked a memory? Maybe the smell of sun cream reminds you of being on a summer holiday? This happens because smells can **trigger memories** that may have been locked away in our brains for a long time.

This experience was first mentioned by a French writer called Marcel Proust, who noticed how when he tasted a particular madeleine cake, certain childhood memories came flooding back to him.

Therefore, this is often known as the Proust effect. Long-term memories are formed when neurons are connected all over the brain. The ones that remember smells seem to have the ability to restart the memory circuit and activate other neurons somehow involved in that memory too.

BRAIN HERO

MARCEL PROUST

French author and first to associate memories with the power of smell, now known as the Proust effect.

Hearing

In hearing people, inside the ears is a set of tiny bones and an eardrum that vibrates when a sound is made. Tiny hairs around them sense the vibration and pass an electrical signal to the auditory nerves nearby. The signal travels to the brainstem and thalamus before reaching the primary auditory cortex in the brain. All the distinct sound patterns are generally quickly analyzed to identify what is being heard.

AUDITORY CORTEX

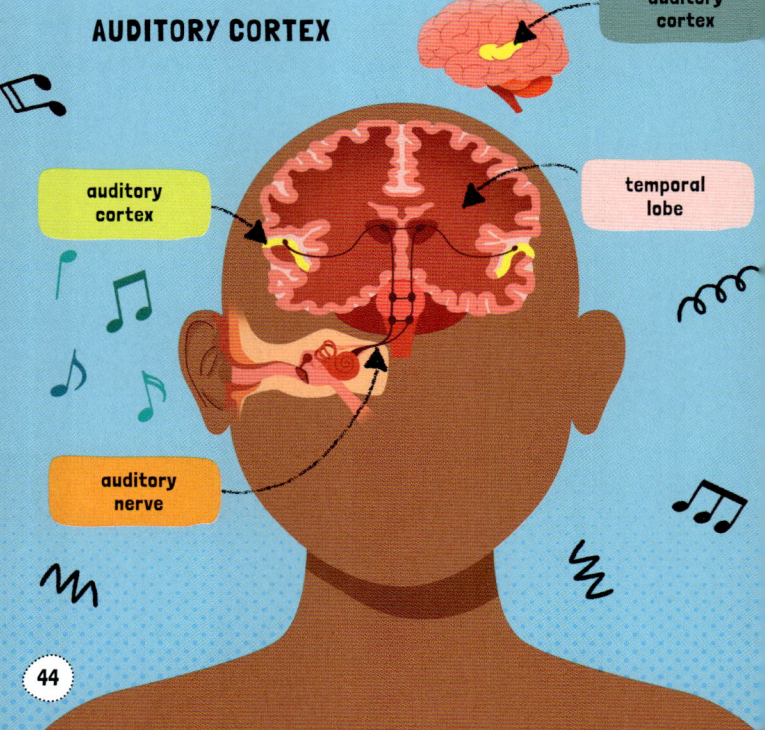

auditory cortex

auditory cortex

temporal lobe

auditory nerve

The brain also adds meaning to the sound. If it is familiar, strange, enjoyable, or frightening, then this is interpreted by the brain, so people know what to do about it. This is important because it might be necessary to quickly understand how to respond, which would be different depending on the type of sound, such as hearing words, a fire alarm, or your favourite music.

Have you ever heard a song that made you feel like dancing, or bursting into tears? Music has amazing potential to provoke strong emotions because it can activate many areas of the brain, including the hippocampus and limbic system, which trigger memories and emotions, and influence how someone feels when they hear it.

Dog ears

Sound moves through the air in waves. Like waves in water, some are fast and powerful (high frequency), and some are gentle and calming (low frequency). Dogs can hear high-frequency sounds (up to around 60,000 Hertz), but humans can't (stopping at around 20,000 Hertz when young, and getting much lower as they age).

Touch

Nerve endings in the skin alert people anytime they touch something. The signal travels along sensory neurons, through the spinal cord, and to the parietal lobe in the brain (see page 28). Here, the brain decodes the information, such as the temperature and texture of what is being touched. This is especially important when touching things that are very hot or very cold, which could cause injury.

Still too hot to drink!

EXPERIMENT TIME

1. If you are comfortable, ask a family member or friend to put the tips of their fingers on your back, but don't let them tell you how many fingers. You have to guess!

2. Now close your eyes and ask that person to do the same thing, but instead of your back, they must lightly touch your face this time. Remember, keep your eyes closed. Can you guess how many fingers they touched your face with? Was it easier this time?

46

Let's see why there is a difference. The area of the skin where the fingers touch plays a crucial part here. In some parts of your skin, such as your fingertips and face, the nerve endings sit really close together. No matter where a fingertip touches your face, there will be nerve endings there, ready to tell the brain about its precise location. But when a fingertip touches your back, the nerve endings there are further apart, so your brain has to guess the location of the finger.

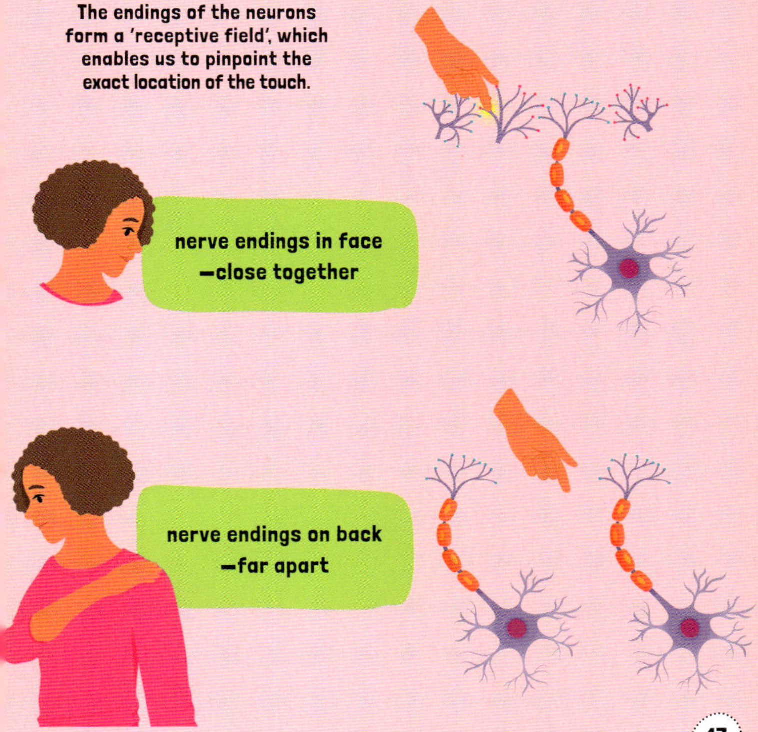

The endings of the neurons form a 'receptive field', which enables us to pinpoint the exact location of the touch.

nerve endings in face
—close together

nerve endings on back
—far apart

Pain

Believe it or not, pain is actually good for us. It alerts us to injuries and is a way for our brain to tell us to take it easy for a while and be careful in the future. Pain signals travel along sensory nerves called **nociceptors** (say no-see-septors). The initial jolt of pain we feel when we bang our head or cut a finger is an electrical signal that travels at up to 140 km per hour along our nociceptors. Once that signal reaches the brain, we become aware of it.

Speak like a scientist

NOCICEPTOR

A nociceptor is a specialized type of neuron that carries pain signals.

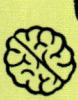

Extra senses

What if we could somehow boost our senses? Today, scientists are busy creating devices called 'brain-computer interfaces', which will be wearable and allow us to control computers and equipment, such as power chairs, with our brains. Right now, these devices can use our thoughts to control the TV, send text messages, control robotic limbs and feel what is touched with them, and communicate with other people (in a very basic way).

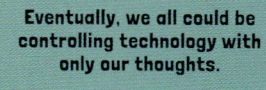

Eventually, we all could be controlling technology with only our thoughts.

In the future, new technology will enable people to see the world with augmented reality, heighten sense of smell, visualize dreams and even share them with other people, and store memories for ever.

Speaking of memories, read the next chapter to find out how your brain learns new things to **create and store unlimited memories.**

Chapter 6

Learning and Memory

Our memories are created in a place called the hippocampus (see page 28). It is essential for our short-term memories, and it is where our brain starts the memory process for things we remember in our lives.

To remember things for longer, the neurons in the hippocampus reach out to other parts of the brain and make a connection with neurons in areas that process different types of information. Each part will remember what we saw, how we felt, who was with us, and what we did so that the neurons in all the other parts of the brain create an extensive network of memories.

For example, when we experience something particularly **big and meaningful,** our brains will associate that event with all of the emotions and details of the day.

In 1953, an American named Henry Molaison, referred to as 'patient H.M.', underwent brain surgery at the age of twenty-seven for severe **epileptic seizures**. While he kept a lot of his memories from before the operation, he lost his ability to remember things that happened more than a few minutes ago. Later, he was willing to help scientists like Brenda Milner better understand how memories work in the human brain. Milner also went on to study other brains, and explain how the brain uses language.

BRAIN HERO

BRENDA MILNER

British-Canadian neuropsychologist who researched how our brain has specific areas for learning and memory.

Neuroplasticity

The brain will always become more skilled at an activity the more effort and practice you put into something.

When we learn anything new, such as riding a bike, the neurons in the brain will reach out and form connections with other neurons. This process of change within the brain is called **neuroplasticity**. Sometimes it is described as the brain 'rewiring' itself. The brain will keep working on these connections over time, especially if you do the new activity often.

The more often you practise or study, the more the brain sees this information as relevant and it makes sure those connections between neurons are working efficiently. When we don't practise something, those connections eventually get weaker and the information or skills become harder to remember over time. This is what neuroplasticity is all about. A scientist called Marian Diamond spent much of her career discovering how neuroplasticity works, and was the first person to explain the process.

Whenever we learn something new, neurons form connections with others all over the brain. Like knowing a frequently travelled route, these connections form pathways that help us to remember what we've learned.

Speak like a scientist

NEUROPLASTICITY

Neuroplasticity is the term given to changes made in the brain when we try new activities, helping it to learn and improve performance.

BRAIN HERO

MARIAN DIAMOND

American scientist and one of the first people to explain the process of neuroplasticity. She also got the chance to study the brain of the world-famous physicist, Albert Einstein.

Types of learning

Whilst a classroom may be an obvious place of learning, our brains also learn things automatically when we don't have voluntary control over them.

One type of automatic learning is called **conditioned learning**. This is when your brain learns to do something because it links a **stimulus** (something that makes you react, such as a sound or colourful light) with an outcome.

A famous example of this would be when a Russian scientist called Ivan Pavlov used sounds every time he fed some dogs. Over time, the dogs would get hungry just by hearing the sounds. This is called classical conditioning.

Another kind of learning is called **operant conditioning**.

 ✳ **Speak** like a **scientist** ✳

OPERANT CONDITIONING

Operant conditioning is a type of learning in which we decide to do things that will bring us rewards and avoid behaving in ways that will be bad for us.

We can use operant conditioning when we train dogs to do tricks like 'sit'. When they sit down, we can give them a treat. Because dogs like treats, they learn to do the trick. They will think about it, and decide to do the trick in order to get the reward.

Improving our learning skills

The more time we dedicate to learning something, such as a new language, the more the process of neuroplasticity occurs. The result is a brain that works more efficiently to remember the skills, experiences, and information that will be useful to us.

The first step to learning is to **repeat something**. By going over the same facts and information, we tell the neurons involved to keep sending signals because it is important that our brains learn it. Over time, it will become easier to remember.

Ask someone to test your knowledge. By doing this, your brain not only activates the memory of what you have learned, but keeps it fresh in your mind, ready to be used for the next time. In fact, when we learn new things we often consolidate our memory of them in our dreams, which helps us better remember our new skills and knowledge.

What's your name? = ¿Cómo te llamas?

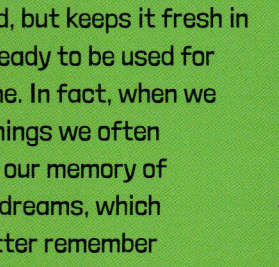

Eating food that will nourish our bodies and brains, and staying hydrated, will all help our learning. Also, try rewarding yourself by noticing when you feel more positive or confident. Giving ourselves a reward tells our brains that learning is a good thing.

Thanks to the work of two Russian neurologists, Alexei Kozhevnikov and Sergey Korsakov, we understand much more about our brain's capacity for learning.

BRAIN HEROES

ALEXEI KOZHEVNIKOV AND SERGEY KORSAKOV

Russian neurologists who specialized in studying the brains of medical professors, artists, scholars, writers, and politicians.

In the next chapter, we'll discover that one of the best things we can do to help our memories improve is **to get enough sleep.**

Chapter 7

Why the Brain Needs to Sleep

Sleep helps our body and brain to recover from the day's activities.

Sleep flushes CSF through the brain, replenishes depleted neurotransmitters, and allows our brain to process new experiences we had during the day, finding a place to store them in our memories. Overall, sleep is critical in helping our brain to function at its best. The Russian scientist and pioneering sleep expert, Maria Manasseina, explained that sleep was even more important than food for a healthy life.

BRAIN HERO

MARIA MANASSEINA

Russian scientist and early expert in sleep, who wrote a book on the subject in 1889.

Circadian rhythm

We generally sleep at night and are awake during the day, but how does the brain know when it is time for bed and when it is time to be awake? Well, our brains are set on a specific twenty-four-hour sleep-wake cycle called the **circadian rhythm** (say sir-kay-dee-an).

The brain has its own internal clock controlled by things like sunlight, temperature, food intake, and levels of a **hormone** called **melatonin**. This clock is in a part of your brain called the hypothalamus, which has lots of functions, from controlling your appetite to regulating body temperature and even helping to create your emotions. However, the hypothalamus also contains a group of brain cells that form the SCN (**suprachiasmatic nucleus**), which functions as a clock, ensuring our body and brain follow a sleep-wake cycle.

Speak like a scientist

HYPOTHALAMUS

The hypothalamus is a small but essential region involved in appetite and being thirsty, sleep, body temperature, and releasing hormones.

How much sleep do we need?

The amount of sleep you need throughout your lifetime changes, so your brain constantly adjusts its clock. For example, babies require more sleep because everything is brand new to them, so their brains need extra time to process the world around them. By the time we are around sixty-five years old, the SCN doesn't work as efficiently as it used to, affecting our sleeping patterns.

0–18 MONTHS

11–17 HOURS

18 MONTHS–2 YEARS

11–14 HOURS

3–8 YEARS	9–13 YEARS	14–17 YEARS
10–12 HOURS	**10 HOURS**	**8–10 HOURS**

18–64 YEARS	65+ YEARS
7–9 HOURS	**7–8 HOURS** (But this may be restless sleep.)

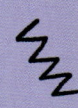

Light and deep sleep

There are two main types of sleep—light and deep sleep. The brain cycles through them several times every night, going back and forth between the two. The different stages of sleep can cause our eyes to move rapidly (even though they are closed), especially when we are dreaming. That is why light sleep is called Non-Rapid-Eye-Movement (NREM) sleep because our eyes don't move much, and deep sleep is called Rapid-Eye-Movement (REM) sleep because our eyes move a lot.

61

Your **brainwaves** (signals from the active neurons throughout the entire brain) also change greatly during sleep. There are five stages of sleep, from being awake and alert to being tired, drowsy, NREM sleep, and finally REM sleep, and they all have different brainwave activity.

 ★ **Speak** like a **scientist** ★

BRAINWAVES

Brainwaves is the term for the collection of signals from all over the brain that are measured as waves of signals, rather than signals from individual neurons.

Dreams and nightmares

The brain experiences dreams during sleep as it processes and organizes everything we've done during the day. When we sleep, the brain strengthens memories from the previous few days so that they stick in our long-term memory storage. Not getting enough sleep can limit this process.

The brain doesn't go to sleep completely when we do. The visual cortex (within the occipital lobe) is very active and can allow our dreams to have visual aspects to them. But because the frontal lobe is sleeping, our dreams can often feel strange and confusing. When we are asleep, the brain doesn't have to use logical decision-making and context, like it usually does when we are awake. It is free to dream incredible things.

The brain experiences nightmares in a similar way to dreams, but in our nightmares, an additional part of the brain is awake, called the amygdala (say ah-mig-dala). The amygdala is essential in helping our brain feel all of our emotions, but it is particularly important in fear.

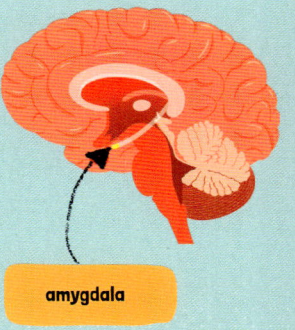

amygdala

Speak like a scientist

AMYGDALA

The amygdala is a small region of the brain that is essential for all emotions, but especially fear and anxiety.

The amygdala in particular is really active during nightmares. We might get nightmares when we feel anxious or worried during the day, and it transfers over into our sleep.

Hypnosis

Hypnosis is a technique that puts someone into an **altered state** somewhere between being awake and asleep. Even though it has been around for hundreds of years, probably longer, scientists don't really understand what is happening in the brain during hypnosis.

Under the careful guidance of a trained professional, people can fall into a very relaxed state and are much more open to suggestions and instructions. People can use hypnosis to reduce **fears** and **phobias** or remember lost memories. Some people can even undergo surgery when they are hypnotized, rather than using anaesthetics to put them to sleep.

Sleepwalking and sleep talking

Sometimes, people can be sound asleep but still manage to get out of bed to move around or talk, even though their brain is sleeping—well, at least part of it is. Usually, when we sleep, our brainstem sends out neurotransmitters like GABA (see page 13) to instruct neurons to stay quiet, especially ones involved in movement. This stops us from moving around so that we don't act out our dreams. When people sleepwalk, it's because some parts of the brain stay more awake, particularly in the brainstem, cerebellum, and motor cortex (responsible for movement), meaning they can move around even when asleep.

Sleepwalking is very common, with about 30% of children doing it at least once, and sleep talking is even more common. If you want to learn about how our brains talk and use language, **carry on to the next chapter.**

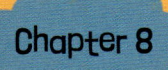

Chapter 8

Speech and Language

We communicate with other people every day using different forms of language, such as speech or sign language.

When someone communicates with us, our brains try to understand what they are telling us. When we want to respond, an area called the pre-motor cortex will prepare the signals to our mouths or hands, giving instructions for how to respond.

Brain regions

One unique thing about language is that, unlike other processes in your brain—which tend to use both

BRAIN HERO

PIERRE PAUL BROCA

French physician who identified the part of the brain responsible for all forms of language.

hemispheres—all forms of language rely mostly on the left hemisphere.

When you use language, an area of the parietal lobe called the supramarginal gyrus works like a conductor of an orchestra, activating and synchronizing many different brain regions to help you interpret the meaning behind what is being communicated. **Sign language** activates the parietal lobe more than spoken language, as it becomes more involved in the complex hand movements.

In 1861, Pierre Paul Broca realized that a specific area of the brain helps us to understand language and how sentences should be constructed. We call this Broca's area and now understand that it is vital in many types of language, such as spoken and sign language.

In 1874, Carl Wernicke discovered there is a specific area in the brain that helps us understand the meaning behind words.

BRAIN HERO

CARL WERNICKE

German physician who identified the part of the brain responsible for understanding the meaning of words.

Body language

When we have a conversation with someone, most of the information our brain receives doesn't come from words—it comes from how the person is speaking.

If a person is yelling at us because they are angry, or whispering because they are afraid, our brains need to detect that so we can respond appropriately.

As well as spoken language, the brain needs to be able to detect and interpret **body language**. About 60% of language is non-verbal, meaning it is not spoken. Similarly with sign language, a large proportion of the meaning comes from the gestures, movement, and facial expressions that are used alongside the signs. In general, similar regions in the brain are used to interpret the meaning behind verbal or non-verbal communication.

Body language (anything that we do with our body, hands, or facial expressions when we talk) also helps to support what we want to communicate, which is especially useful in situations where we don't want to (or can't) use words.

The emotional parts of our brains, connected together in the limbic system, are also activated to allow us to feel a similar emotion to the other person (this is called empathy, the ability to understand someone else's perspective).

Understanding body language is something that the brain improves on over time. If you want to know how we understand so much about how the brain works, **then read on to the next chapter.**

Chapter 9

How Do We Study the Brain?

With all this talk about the brain, you might wonder how we know so much about it. Well, to find out how it works, we study it. But, as people tend not to enjoy having their brains scooped out for examination, we have had to find better ways to study them. Let's take a look at how it's done.

CT scan

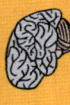

If you get ill or hurt your head, a doctor might want to check out your brain to ensure there are no serious issues. They use sophisticated cameras to examine your brain.

A computerized tomography or CT scan is one way to do this. A person lies on a bed inside a giant machine that looks like a big doughnut. It takes lots of **X-rays** of their brain from all different angles to produce a three-dimensional image of their head and brain.

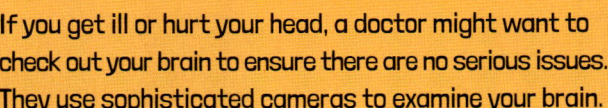

MRI scan

Another way to look at the brain is with MRI scans, or magnetic resonance imaging, giving a sharper and more detailed image. The MRI machine generates a **magnetic field** (thousands of times more powerful than the Earth's magnetic field) that interacts with **hydrogen atoms** in your body. The machine detects this interaction to create a picture of the brain.

The MRI is so sophisticated that it can look at other parts of the body, not just the brain. It takes many thin pictures and adds them together, one layer at a time, to create a detailed image.

Can you identify any brain structures, such as the brainstem, on this MRI scan? The top row shows a side view of the brain and the bottom row shows a front-facing view.

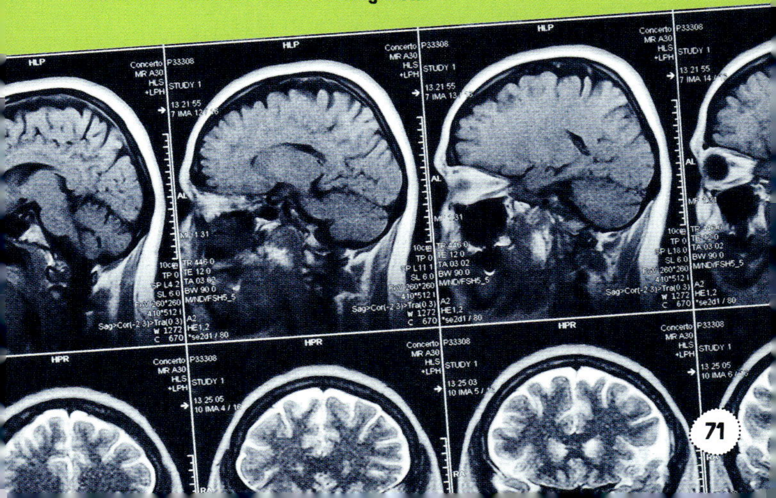

71

The functional MRI

But what happens if we want to see the brain in action? MRI scans can still be used, but now they get an upgrade. When the brain sends signals, it uses up a tiny bit of oxygen. The functional MRI (fMRI) looks at the oxygen atoms and records the oxygen change when your brain does something, like thinking of a memory or moving your hand. The fMRI shows us what parts of the brain are 'thinking'. In the image below, the colours represent brain activity and increased oxygen in the areas that are working hard.

THE MOST ACTIVE AREAS OF THE BRAIN

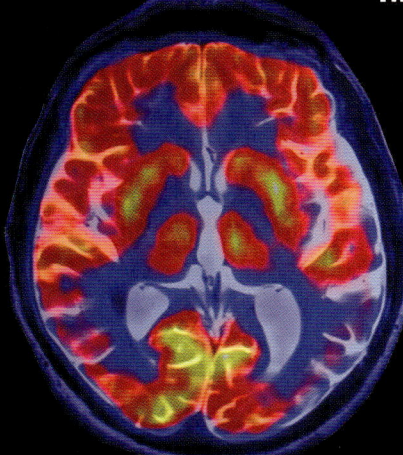

fMRI scan of the brain, as seen from above the head looking down. The colours come from a chemical tracer (something that is designed to be seen) going to areas of the brain that are more active.

In the late 19th century, Italian physiologist Angelo Mosso developed a way of measuring changes in blood flow in the brain. Called the human circulation balance, it measured the weight of a person's head and showed that blood flow in the brain changes when we use it more. It is the basis of how we use fMRI machines today.

BRAIN HERO

ANGELO MOSSO

Italian physiologist who studied blood flow in the brain and developed a way of measuring it.

Brainwaves can also be used to examine brain activity. We already know that brainwaves are waves of signals from all of your neurons at the same time. Instead of cameras, doctors can place a set of **electrodes** on to the head. Electrodes are small devices that conduct tiny amounts of electricity that can be measured. They are capable of detecting how active the brain is by measuring brainwaves. This is called an EEG, or electroencephalogram (say electro–en–kefalo–gram).

What do neuroscientists do?

Neuroscientists study the brain to learn more about it, hoping to create new medicines and treatments. One way they like to explore the brain is by examining tiny pieces under a powerful microscope. Some of these come from patients undergoing brain surgery, or they are grown from single brain cells in a laboratory.

Scientists can see what the brain cells are doing and run experiments in laboratories by thinking up a **hypothesis** and testing it in fun and interesting ways to see if they were right. This is how new scientific discoveries are made.

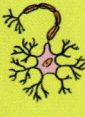

Scientists can even put brain cells into a **Petri dish** in the laboratory, where new ones will stay alive and grow. A dish full of neurons can be tested in many different ways. For example, potential new medications can be tried out on them to see how the neurons respond.

Before there were powerful microscopes to take pictures of neurons, people had to draw them by hand.

Santiago Ramón y Cajal's drawings were similar to this one.

Nobel Prize-winning neuroscientist Santiago Ramón y Cajal produced highly accurate drawings of the anatomy of the brain.

Remember how brain cells like to communicate with each other and form connections? The action potentials (see page 17) that travel along the neuron can be observed by scientists using tiny specialized electrodes. Scientists can also make the neuron change colour during an action potential so they can see it.

BRAIN HERO

SANTIAGO RAMÓN Y CAJAL

Spanish neuroscientist working in the late 19th century. His detailed drawings of neurons are highly accurate when compared to today's microscopic images.

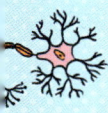

What do psychologists do?

Sometimes, just communicating with a person can help us discover what the brain is really up to. Psychologists and therapists study how the brain affects our behaviour. Often, people have feelings that are complicated and difficult to understand. By asking questions about how a person feels, they can help someone cope with things like depression and emotional challenges.

Although not all of his ideas have been correct, early 20th century psychologist Sigmund Freud is famous for his studies into the minds of people, and introducing the idea that there may be a reason why we behave the way that we do in certain situations.

How are you feeling?

BRAIN HERO

SIGMUND FREUD

Austrian psychologist who invented psychoanalysis, a method for analyzing our minds to understand our behaviour.

You might even see these psychologists in schools where they use tests and questionnaires to understand a student's brain a little better. But don't worry, these are not the kind of tests you need to study for. These 'aptitude tests' are designed to see how a student's brain uses memory, logic, planning skills, reading, and listening skills, so psychologists can figure out how to help the student build on their skills.

For many years it was thought that brains should all work in the same way. However, more recent discoveries have shown us that there are various different ways for a brain to interpret things.

In the next chapters, let's take a look at how brains vary.

Chapter 10

Neurodiversity and Neurodivergence

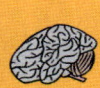

Although many brains function in a similar way, each human brain has slight variations in the way it works. This difference between brains and how they interpret the world is known as neurodiversity.

Neurodivergence is a term used to describe greater variations in the brain from what is seen as average, or **neurotypical**. This divergence is a natural variation.

✳ Speak like a scientist ✳

NEURODIVERSITY

Neurodiversity is the term given to the differences in the way brains work.

NEURODIVERGENCE

Neurodivergence is the term used when the processes within the brain result in patterns of learning or behaviour that are different from 'neurotypical'.

Today, we are much better at describing some of the ways in which brains function. This might lead to distinctive ways of thinking or behaving. Neuroscience is now teaching us that this diversity is much more common than we originally thought—it is just part of how brains develop.

Dyslexia

Dyslexia (say dis-lex-ee-ah) is a term that is used to describe one type of neurodivergence where people show strengths in creative and visual work or problem-solving, but face challenges in reading and writing. It affects 5-20% of all people worldwide.

Challenges in putting details in order

Excellent long-term memory

Challenges in processing and remembering sounds

DYSLEXIA: STRENGTHS AND CHALLENGES

Strong 3D visual skills

Strong communication skills

Challenges in reading and writing

Strong problem-solving

Dyslexia can make reading and writing especially challenging because activity in the areas of the brain involved in those things is disrupted. Dyslexia can also make it hard for the brain to process things that are seen and heard, making it challenging for people to organize all of their thoughts.

Dyscalculia (say dis-cal-kew-lee-ah) is the term for specific challenges in dealing with numbers.

Autism

The autism spectrum represents another neurodivergent condition. It is what is known as neurodevelopmental, which means that changes in the brain occur when it is developing.

The characteristics of autism can vary hugely between people. Really, autism is a term for a vast number of traits that are never quite the same between two people.

Autistic people process information in different ways, which can result in great skills, such as an ability to remember facts and solve problems in new ways. Features of autism can range from finding social situations challenging, or finding sensory stimulation such as sound overwhelming, to finding non-verbal cues such as body language hard to understand.

Attention to detail

Absorb and
retain facts

Deep focus

**AUTISM:
SOME POSITIVES**

Visual skills

Observational skills

Tenacity and
resilience

Accepting of difference

The challenges a person with a neurodivergent brain may experience will also vary depending on the person.

Autism is quite common. About one in a hundred children are diagnosed as autistic, with boys more likely to be diagnosed than girls.

ADHD

Attention-deficit/hyperactivity disorder, or ADHD, is a neurodivergent condition affecting about 5% of all people. While there are many traits that are unique to individuals, they generally fall into a couple of categories, such as challenges in paying attention and planning things, hyperactivity, and being impulsive.

ADHD is typically
diagnosed early in life,
before the age of twelve, but
it is also common for people to be
diagnosed in their late teens, or as
adults, too.

Epilepsy

Epilepsy is a neurodivergent condition where the
brain experiences seizures. Typically, there are
millions of signals firing all over the brain anyway,
but they are tightly controlled by different types
of neurotransmitters. Sometimes, the brain
temporarily loses its tight control of the activity,
resulting in a seizure. Different parts of the brain
can be affected, depending on where the seizure
occurs. Seizures can occur spontaneously, or they
can be triggered by things like strobe lighting
or loud noises.

There are many different types of seizure,
but they can cause someone to shake
or become very stiff and rigid, or be
unresponsive.

People with epilepsy often don't remember the seizure afterwards.

In most cases, doctors can manage seizures with medicines that are able to reduce their frequency, although the side effects of medications can also be undesirable and leave a person feeling tired, with difficulty concentrating.

Epileptic seizures don't generally damage the brain, even if they do affect someone's daily life. But they can be dangerous and lead to physical injury, for example, if someone is standing and they fall or if they are driving a car.

With the right medication, epilepsy can have minimal impact on a person's day to day life, and they often go years without experiencing any seizures at all.

Sometimes, changes occuring in the brain present challenges that result in a significant effect on someone's daily life. In the next chapter, we will describe some of those changes and what they mean for the brain.

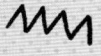

Chapter 11

Brain Health

Brain health can be impacted by various things, and in various ways, some of which we're going to explore in this chapter. We'll also look at how we can look after our brains to maintain their health.

Dementia

Neurodivergence is about how the brain is wired, but things can also impact the brain at various points in life that can cause it to work differently. Neurological changes in the brain can have a big impact on someone's life.

One example of this is dementia. Dementia is a term we use to describe conditions when a brain loses some of its functions, like memory, social skills, and thinking skills. Dementia can result in damage to neurons that would normally send signals throughout the brain. Therefore, communication between neurons in the brain can be affected, resulting in a loss of brain function. The most common areas affected are language and memory, making it difficult to think clearly.

The risk of dementia increases with age, which is why it is mostly seen in people over the age of eighty-five. The most common form is Alzheimer's disease. While we still don't know exactly what causes it, and unfortunately it isn't something that gets better on its own, current research is working hard to find more effective treatments and ways to prevent it.

Let's listen to some of your favourite songs, Nana.

Cerebral palsy

Low levels of oxygen, head injury, bleeding in the brain, or certain types of infection either before or shortly after birth can lead to cerebral palsy. It affects around 0.2% of the population and impacts the messages between the brain and the body, affecting movement and communication in various and differing ways, depending on how the brain was affected. If people with cerebral palsy need support in communicating and moving around, various equipment is available, alongside physical and speech therapy.

Stroke

Damage to part of the brain can occur when too little oxygen gets to the brain cells. For example, this can be caused by a stroke. An ischemic (say iss-kee-mik) stroke happens when blood vessels in the brain get blocked and fresh blood, containing oxygen, cannot get to a certain region. A hemorrhagic stroke occurs when blood vessels burst open and spill their blood. During a stroke, blood isn't able to get to where it needs to go, and brain cells quickly die or become damaged.

Head injury

Head injuries can also lead to bleeding and swelling in the brain, causing neurons to die. This can affect personality, memory, movement, taste, vision, and many other things. The best way to **protect our brains** is to wear a helmet if we do anything that could cause us to hit our heads, like rollerblading or cycling.

Even playing sports where we need to use our heads, like football, can harm the brain. Doctors and neuroscientists are just now realizing that using our heads for sports can damage our brain if we do it over many years. So try and be as careful as you can. Your brain will thank you for it.

Drugs and alcohol

Alcohol and illegal drugs can also have severe side effects on your brain. They alter how the neurons communicate with

each other, can change how awake or sleepy we are, can make us feel sad and depressed, and can make us do things we otherwise wouldn't want to do.

Looking after your brain

What you eat is crucial for your body and your brain. Some foods are more beneficial than others, so it's important to eat a good range, including those that supply vitamins, nutrients, and healthy fats. It's also vital to drink plenty of water. Choosing foods in a range of colours that are in their natural state—including vegetables, grains, proteins and fats—will help your body get what it needs.

Because neurons have fat in their membrane, you need to eat healthy fats to keep them working well. Omega-3 fats found in fish, avocado, olive oil, nuts, and seeds are great for your brain and have been shown to keep it healthy for your whole life. And, of course, staying hydrated is important too.

Exercise your brain

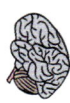

The good news is, the brain doesn't really need much to keep it running well. One of the best things you can do to keep your brain healthy is to exercise, even if it is only a little bit each day. Keeping physically active helps your brain to stay healthy. Finding something you enjoy means you are more likely to do it. It keeps neurons working at their best, and can help to protect the brain, especially as you get older.

89

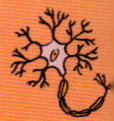

A brain for life

Hopefully, by now you have learned how incredibly sophisticated and brilliant your brain is. This **big computer inside your head** is always trying to learn more and gain fresh insights about life by experiencing new things every day. The best thing is, your brain will do this every day, all on its own, you just need to keep it healthy and happy. You and your amazing brain have a lifetime of adventure and learning to get through, and now you get to decide how you can do it!

Find things that make you feel good

Spending time with other people and doing things you enjoy with them makes you feel good, thanks to neurotransmitters like dopamine and **oxytocin**. Not only that, but doing things with others helps your brain learn important social and communication skills, which it will keep improving for the rest of your life. Playing things like board games or computer games is also a great way to test your brain and improve your problem-solving skills.

Reading also benefits the language centres of the brain, as does listening to audio books or podcasts.

Of course, you have already read this book and learned so many new things, but now it is time to get out there and find new and interesting ways to challenge yourself. **Your brain will love you for it.**

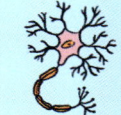

Glossary

action potential the electrical signal that is sent along the axon of a neuron

astrocyte a type of brain cell that helps neurons do their job and keeps the blood-brain barrier working well

blood-brain barrier cells that line blood vessels within the brain and only let beneficial things in, keeping damaging things out

body language anything that we do with our body, hands, or eye contact, that conveys meaning about our feelings or intentions, without being spoken using words

brainwave a collection of signals in the brain, sent by neurons, that can be measured all together

cells the individual building blocks that make up the body. Each one has a specific role to play. Examples include a red blood cell, a skin cell, and a neuron

circadian rhythm the twenty-four-hour cycle of sleeping and being awake

conditioned learning a type of learning that occurs when we link a behaviour or stimulus with a specific reaction

corpus callosum a structure in the middle of the brain that connects the two hemispheres, enabling communication between them

cranial nerves a collection of twelve paired nerves that send messages to the brain, mostly about what our senses are detecting

DNA (DeoxyriboNucleic Acid) the genetic code needed to make every type of cell in our body

electrodes small devices that carry an electric current

endothelial cells cells that line the inside of blood vessels, important in forming the blood-brain barrier

epileptic seizure a common symptom of epilepsy, where the brain signals are disrupted, temporarily making the body shake or become stiff, and unresponsive

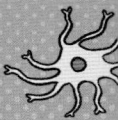

hemispheres the two halves of the brain

hormone a chemical that is used as a messenger to other cells in the body or brain, released into the bloodstream to travel to its target location

hydrogen atom the most basic form of a typical atom, comprising of only one proton and one electron

hypothesis an idea that is tested in a scientific experiment

ion an atom or molecule with an electrical charge

limbic system a group of brain regions responsible for generating our feelings and emotions

lobe a specific region of the brain where certain tasks are perfomed

longitudinal fissure the deep ridge between the two brain hemispheres as seen when looking down onto the brain

magnetic field a field produced by electrical charges of particles that can attract or repel other magnetic particles

melatonin a hormone that tells your brain when it is time to sleep

myelin a fatty substance that coats axons on certain neurons to make them more efficient at sending action potentials

neurodivergence when the processes within the brain result in patterns of learning or behaviour that are different from 'neurotypical'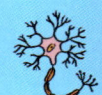

neurodiversity the understanding that all brains are slightly different and interpret the world around us in their own way

neuron a type of cell that sends messages within the brain and throughout the body. A bundle of neurons is called a nerve

neuroplasticity the term for changes in the brain that result from learning and developing

neurotransmitter a small messenger chemical that passes from one neuron to another

neurotypical describes brain development and function that is considered to be within the typical range for a human being

nociceptor a specialized type of neuron that transmits pain signals

operant conditioning a type of learning when we decide to do things that will give us rewards and try not to do something that will be bad for us

oxytocin a hormone produced in the hypothalamus that acts as a chemical messenger in the brain, important in love and bonding

Petri dish a small plate which can be filled with nutrients that cells can grow in, often found in laboratories

psychoanalysis a type of treatment where the inner mind is brought to the attention of the person in an effort to help them understand challenges they face

sign language a language, particularly used among deaf people, that makes use of hand, face, and other body movements as ways of communicating

stimulus something that can cause a reaction, like a loud sound, bright light, or strong vibration

suprachiasmatic nucleus a small bundle of cells in the hypothalamus that has a very important role in the sleep–wake cycle

thalamus a region in the middle of the brain that is important in regulating signals about your senses being sent between the brain and the body

X-ray energy emitted from electrons. Because this energy passes through our tissue more easily than our bones, they produce a shadow of our skeletons, which doctors can then examine

Index

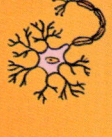